UNIT STUDIES IN SCIENCE

General Editor: D. Shires

Biology Unit Five

Evolution and Heredity

F. G. Wight

DRAWINGS BY M. ABBOTT

EDWARD ARNOLD

© F. G. Wight 1972
First published 1972 by
Edward Arnold (Publishers) Ltd.,
25 Hill Street, London W1X 8LL

ISBN 0 7131 1729 X

Other *Biology* Units:
1. Plant Physiology
2. Animal Physiology
3. Animal Classification
4. Plant Classification

Physics and *Chemistry* Units are
also available.

Printed by Straker Brothers Ltd., Whitstable

Contents

Acknowledgements

We would like to thank the following for permission to reproduce copyright material (page numbers in brackets):

J. Allan Cash (13, 14)

Australian News and Information Bureau (40)

Bruce Coleman Ltd. (4)

Dominion Museum, Wellington, New Zealand (9)

Faber & Faber Ltd. and the Danish National Museum (cover and 16)

Field Museum of Natural History, Chicago (28 top, 44)

French Government Tourist Office (45)

High Commissioner for New Zealand (37)

L. F. La Cour, John Innes Institute (52)

A. Manktelow: *Life Before Man* by A. & C. Black Ltd. (27)

Marine Biological Association of the U.K. (27)

National Coal Board (17)

Natural History Museum (4, 15, 23, 26, 28 bottom, 29, 30, 32–3, 34, 36, 38, 39)

Palaeozoological Institute, Polish Academy of Sciences, Warsaw (18)

John Topham Ltd. (55, 56 top and middle)

Viking Press, New York (35)

Zoological Society of London (20, 42)

Introduction to the Series

It is in some senses a contradiction in terms to write of "text-books" for the Certificate of Secondary Education. Even allowing for the diversity of the Mode One syllabuses published by the various Regional Examination Boards, it would be difficult to estimate the precise scope and intentions of the multiplicity of Mode Two and Mode Three syllabuses which have been produced by individual schools. It is also inevitable, and desirable, in such a flexible examination system that the content of and approach to school work in science will show many changes as freedom to experiment with new ideas shows rewarding achievements in new directions.

With this in mind it was felt that the needs of teachers were not likely to be met fully by text-books of the conventional kind, and that a new approach should be made to the problem. The best solution seemed to be to offer a series of small volumes from which the individual teacher might select those which, in combination, most nearly fulfil his particular needs without including much unnecessary material.

Some twenty such small volumes or units have been planned, grouped into three sets dealing respectively with chosen topics in Biology, Chemistry and Physics. The entire project has been planned so that combinations drawn from these twenty units should serve the needs of pupils studying General Science, or Physics with Chemistry, in addition to those studying the separate science subjects.

Each unit takes the form of a short monograph, written by one or more practising teachers. While adequately covering the subject matter considered appropriate in its own field, it is hoped that by the use of a continuous narrative form each unit will suggest an approach to the topic as a whole which is suitable for pupils preparing for C.S.E. science examinations. Needless to say further units will be added to the series when new requirements become apparent and in this way it is hoped that the series will remain responsive to the flexibility inherent in the organization of the C.S.E. Suggestions from teachers to this end will be very welcome.

The volumes in this series are intended for use at some time during the two years or so prior to the C.S.E. examination, and are written with the assumption that the more elementary aspects of the subjects concerned will have been covered in an introductory science course such as that in the two volumes of "Science through Experiment" to which the present series is intended to be complementary. Used in this way each unit in the series should provide sufficient material to occupy the major part of a school term.

Appreciation must be expressed to the publishers of the series for many helpful suggestions and unfailing patience during the planning stages of the project; and to the authors concerned for the efforts they have made to make their own contributions reflect the overall concept.

D. Shires.

General Editor.

1 The Theories of Evolution

From very early times, naturalists have wondered how the plants and animals which populate the earth in such infinite variety have come to be here. Long before the Bible was ever written, the ancient Greeks suggested that the plants and animals had developed by a system of change from already existing types. This was pure guess work as they had neither the ability nor the knowledge to prove or disprove their ideas. It was not until the nineteenth century that Charles Darwin provided evidence which showed that they were right.

Various theories have been popular from time to time. One which had many supporters right up to the time of Darwin was that the events of the Creation as described in the book of Genesis were literally true. It was known as the *Theory of Special Creation*. Briefly it was that all plants and animals were created as they are now.

Another nineteenth-century theory was the *Theory of Catastrophism* in which the idea was that various regions of the earth have suffered from a series of catastrophes by which everything living was wiped out to be replaced by a new plant and animal population which remained unchanged until the next disaster. This was supposed to have happened some twenty-seven times and each time new living organisms appeared from no one knew where. The presence of fossils in the rocks was explained by saying that they were the remains of earlier populations which had no connection with the new.

The *Theory of Organic Evolution* is very different from the earlier theories. It states that the plants and animals of today are the descendants of earlier types which have changed to suit changing conditions. This last theory is, without doubt, the most acceptable and is supported by nearly all the evidence which we have at present. The two men whose ideas have had most influence on modern thoughts about evolution are Jean Baptiste Lamarck (1744–1829) and Charles Darwin (1809–1882). Both accepted the theory of organic evolution but gave different explanations as to the way in which it happened.

Lamarck began training for the priesthood — a calling for which he had little liking or inclination. His real desire was to be a soldier and at about seventeen years of age, he left college and joined the army. An accident resulted in his being discharged and he began to study medicine in Paris. His interest in this soon changed and he became in turn a professor of botany and later a student of zoology.

Both as a botanist and a zoologist his early views on evolution were those which were popular at the time. These said that a species of plant or animal could not be changed in any way. Later, he revised his opinion and accepted the fact that changes do take place. To explain why the changes occur he said that if, in order to meet the needs of its environment, an animal has to use some structures more than usual and as a result, these structures become changed, then the changes would be handed on to its offspring. Similarly, structures not used would be inclined to become reduced and the offspring would inherit the reduced features. In support of his ideas Lamarck quoted the cases of the giraffe and the flightless birds like the ostrich. The former stretched their necks to reach the foliage on high trees and the long necks which resulted were handed on to the offspring. The flightless birds on the other hand do not use their wings for flying and so successive generations have smaller and

Fig. I   Giraffes have developed long necks.

smaller wings. Briefly, Lamarck said that animals can inherit features developed by their parents. His ideas are not supported by evidence which is available at the moment.

In contrast to Lamarck, **Darwin** was, if this is possible, born a scientist as he was one of a family noted for their scientific ability. His father, his grandfather and three of his sons were Fellows of the Royal Society — an honour granted only to the most eminent. Like Lamarck, Darwin's early scientific training was in the field of medicine but, he soon changed to study for the church. During his time as a student, he developed an interest in nature and became such an expert biologist that he was appointed naturalist on a government survey ship called the Beagle. During the five year voyage, he visited the Galapagos Islands in the Pacific Ocean. Here there is a tremendous variety of plants and animals which has developed in complete isolation. The various types there have often developed from a single species and from what was originally one type of say a bird, there have developed many very different types.

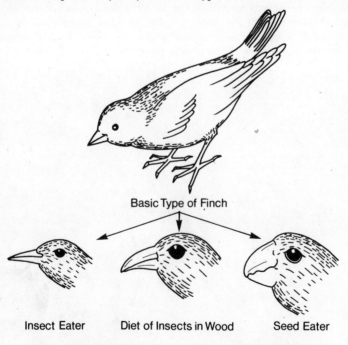

Basic Type of Finch

Insect Eater   Diet of Insects in Wood   Seed Eater

Fig. 2 Darwin's finches — these are some of the types, all of which have developed from a common ancestor.

Darwin thought a great deal about the strange animals of the Galapagos and for twenty years he collected evidence about them and about plants and animals in similar situations. Finally, in 1859, he published his most famous book — *The Origin of Species* in which he outlined a new theory of evolution. The year before this, he had received a letter from a naturalist called Wallace in which was outlined a theory of evolution almost exactly the same as his own. The new theory was expounded to the Linnaean Society in the names of Wallace and Darwin.

The *Theory of Natural Selection* or *The Survival of the Fittest* says that although the members of a species are almost exactly alike, they do differ from each other as do the offspring of any pair of parents. Some of these differences will make some of the animals (and plants) more suited to their way of life and their surroundings. These will thrive and survive to hand on their characteristics to their offspring. On the other hand, those with features which make them less suitable for their surroundings and way of life will not survive and so they and their characteristics will disappear from the species. In this way, by the flourishing of the more suitable and the elimination of the less suitable a species can change or a new species can arise. Can you explain the long neck of the giraffe in terms of Darwin's theory?

Fig. 3   A Praying Mantis.

A classic experiment, to demonstrate that the most suitable do survive, uses the Praying Mantis found in Italy. There are two types; one brown in colour found among dead grass and the other green found in live grass. If green insects are placed in brown grass and brown ones in green grass, it is found that after a few days nearly all of both types of insect have been destroyed by birds. On the other hand, records kept of insects in their normal places show that most of both types survived. Obviously, the brown Mantis cannot survive on green grass and vice versa because they are not suited to their surroundings.

Another example which is found in Britain is that of the Peppered Moth. The normal type of this moth has silvery grey wings with dark spots, a colouring which makes it almost invisible against the bark of trees especially if they are covered with lichens. In 1848, a very dark coloured moth of the same type was discovered in East Anglia in areas where the trees were discoloured by the industrial

Fig. 4 Peppered moths.

smoke. Experiments similar to those done with the Mantis showed that the dark coloured moths survived in the smoky areas but not in the country where the trees were clean. The opposite was also found to be true.

It is easy to show that this can happen by cutting out small moth-shaped pieces of coloured paper, about 40 or so from greyish brown paper and a similar number from pale grey paper. These are fastened, some to backgrounds which match them and some to backgrounds which do not. Pupils who have not played any part in the cutting out or in the placing of the "moths" are then asked to find as many as they can in a short time. A brick wall can be used as a background if the "moths" are cut from paper in shades of red and brown.

2 The Evidence for Evolution

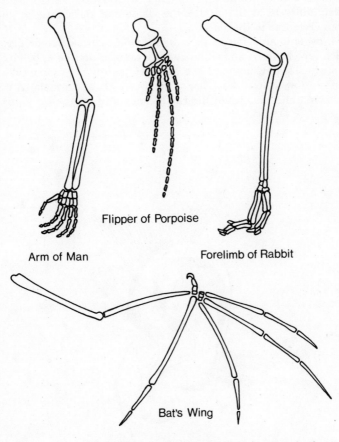

Flipper of Porpoise

Arm of Man

Forelimb of Rabbit

Bat's Wing

Fig. 5 The skeletons of the limbs of different animals show great similarity.

The Evidence from Anatomy

The form and structure of many animals when compared with that
of other animals — the study is called Comparative Anatomy —
reveals many structures so similar that one is forced to accept the
idea that the animals must be descended from common ancestors.
The arm of a man, the leg of a frog, the leg of a tortoise, the wing of a
bird and the wing of a bat all have the same basic structure with the
same bones present. This at least suggests that the ancestors of these
animals must have been very much alike. The legs of horses, deer and
hippopotamus can be seen to be broadly similar to one another and to
the arms and legs of all other animals with backbones. Animals as
widely different as the giraffe, the whale and the mouse all have
seven vertebrae in their necks.

Many animals have in their bodies *vestigial structures* or organs
which now have no known function. Examples are the appendix of
man which is now not used and is often a nuisance. At some time in
man's history it must have had a use, probably one associated with
digestion. The muscles of the ear by means of which the pinna is

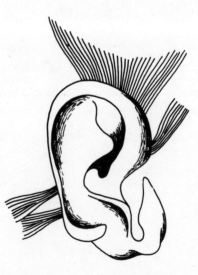

Fig. 6   The muscles which would move a man's ear.

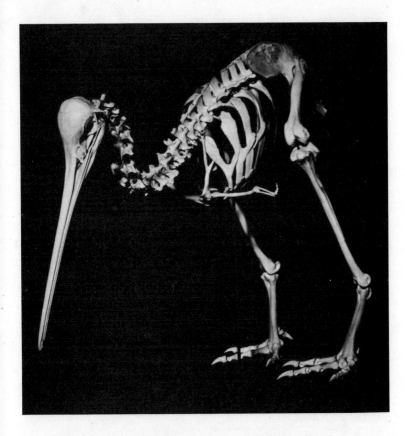

Fig. 7   The skeleton of a Kiwi — a flightless bird from
New Zealand showing its rudimentary wing.

moved in wild animals are present in man but he cannot move the
pinna at all freely. The python and the whale have the bones of very
small hind limbs completely buried in their bodies, a fact which
suggests that their ancestors were animals with limbs.

The Evidence from Embryology

A theory based on the development of animals from their eggs was first suggested by **Haeckel** in the nineteenth century. He said that, during its development, an animal becomes in turn the adult stages of its ancestors. Thus a human baby in its very early stages has five pairs of grooves in its neck which are likened to the gill slits of a fish. Its heart becomes first like that of a fish, then like that of a amphibian, then like that of a reptile and finally like that of an mammal. Although there may be a grain of truth in this theory, it is more likely that an animal, during its development, passes through stages similar to those of the very young of its ancestors. Sometimes, "throw-backs" are born which have vestigial external structures. Examples of this are seen in those human babies which have a very small tail and horses with extra digits on the side of the lower leg.

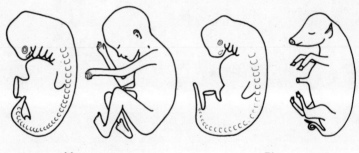

Man   Pig

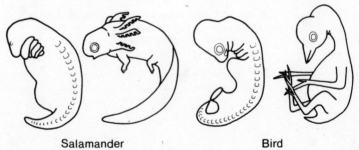

Salamander   Bird

Fig. 8   **The early embryos of different animals look very much alike.**

The Evidence from Distribution

Different regions with similar climates often have very different animals and plants. For example, animals of the North and South polar regions are quite different. We find penguins only in the Antarctic and polar bears only in the Arctic. The Australian animals are different from those found in any other part of the world and those of Africa are different from those of South America. This means that, in spite of very similar climates, the animals have evolved along different lines. Darwin during his voyage on the Beagle also noticed that some animals such as the giant tortoises found in the Galapagos Islands and in certain islands off the coast of India, have become extinct on the nearest mainlands. He also noticed that evolution had, on the same islands, resulted in the development of animals (birds in particular) quite unlike their ancestors and their relatives which were on islands close by.

The Evidence of Convergent Evolution

Another important piece of evidence which shows that animals can change to suit their surroundings is found in the study of what is called *Convergent Evolution*. This is the changing of animals, originally completely different, into similar forms to suit them to their surroundings. Thus, we have as sea-dwellers, both fish and mammals which superficially look alike. The bodies of dolphins, porpoises and whales which are mammals are fish-like in shape and their fore limbs have become like fins. Their hind limbs are much reduced and are buried in their flesh. A now extinct reptile called the Ichthyosaurus was also very fish-like in its appearance. A half way stage is seen in the seals which lead a semi-aquatic life. Their bodies are broadly fish-like in shape and their limbs have become flippers which still have claws.

Plant and Animal Breeding

Later (p. 56) we shall see that evolution, controlled and directed by man, also provides important evidence that it can and does take place.

The Evidence from Fossils

The most impressive evidence of all for evolution comes from a study of fossils — the remains of plants and animals found buried in the rocks which make up the earth's crust.

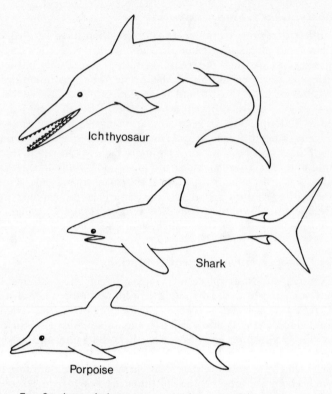

Fig. 9   Animals living in water have similar shapes.
         The Ichthyosaur — an extinct reptile.
         The shark — a fish.
         The porpoise — a mammal

The Formation of Fossils

Let us think first about the formation of the rocks. It is generally believed that the first formed and therefore the oldest rocks are the igneous rocks which began as molten masses which cooled and hardened. The exposed surfaces of these have been attacked by the weather ever since they were formed. Rain, heat and cold, rivers, glaciers and chemical action all played their part and resulted in the rocks being broken up into very fine particles, many as fine as the finest dust. These small pieces were carried away, some by

Fig. 10   The Grand Canyon of the Colorado.

wind but mostly by water, especially by rivers which carried them to a lake or to the sea where they sank to the bottom and became solid to form what the geologist calls sedimentary rocks. In theory, the layers in which they lie should be in order of time and thus form a perfect calendar. Unfortunately, upheavals like earthquakes and the natural slow movements of the crust have upset the pattern and often we find the older rocks on top and the younger rocks beneath. In a few places, the order has not been disturbed. Two such places are the Grand Canyon of the Colorado river and certain areas of the Badlands of Wyoming and these have provided a lot of information about the ages of rocks. It has been calculated that the chalk which is found in our southern counties grew in thickness at the rate of 1 cm each 1000 years. The age of any fossil found in it can therefore be calculated. Rocks can also be dated by studying the radioactivity of the minerals in them.*

Fig. 11   Rock formations showing strata.

* See this series *Chemistry, Unit 1*, pp. 46–51.

When the sedimentary rocks were being formed plants and animals living in those times died. In most cases their remains rotted away or were eaten but some were buried in the mud which then hardened into rock. Mostly it was only the harder parts which were preserved in this way. Fossils can be made in various ways. Sometimes, but very rarely, the actual bones are found. More commonly, the bones are impregnated by mineral matter and literally turned to stone, or petrified. Sometimes the bodies of the animals (or plants) rotted away after they had been encased in the rock and left hollows which acted as moulds which filled with some mineral so that casts

Fig. 12 A fossilised footprint of a Dinosaur.

were left behind. All living things contain carbon and sometimes just the carbon is left behind as a print of the original shape. Any remains are regarded as fossils and a footprint is often of great value in telling us about the creature which made it.

Sometimes animals are preserved intact. In California there are some areas where tarry material is found naturally on the ground in large pits. Animals are caught in the tar in the same way as flies are caught on a flypaper and their bodies are preserved whole. Peat bogs will preserve bodies. An example of this was seen when in 1950 a perfectly preserved man who had been buried for 2000 years was found at Tollund in Denmark. Mammoths, those giant hairy elephants which roamed the earth perhaps 1 000 000 years ago have

Fig. 13   Tollund man.

been found frozen in the ice of Siberia. These were so perfectly preserved that the grass which made up their last meal was still in their mouths.

The Finding of Fossils

Fossils may be found in various ways. They are frequently discovered by accident when mining or quarrying operations are carried out — coal contains many plant fossils — but very often, fossils are exposed when the rocks in which they are buried are worn away by the weather. Sometimes, if a region is discovered which looks as though it might be rich in fossils a special operation is put underway of which the main purpose is the discovery of fossils. A well-known example of this is the "dig" of Dr. Leakey in Africa where a search for human remains has been in progress for many years.

Fig. 14   Fossils in coal.

Whatever form it takes, once a fossil has been found great care and patience is needed to separate it from the rock and often to arrange the pieces in the correct way.

This fossil evidence provides us with a great deal of information about the living things which have inhabited the earth, but it must be remembered that to make sense of what is found is like trying to read a book from which most of the words have faded, from which many pages are missing and those which remain are torn into very small pieces and well mixed up.

Although the records left by most of the animals is incomplete, there are some few cases where the entire story has been worked out — the horse is one of these. Remains of the earliest ancestors of the horse were found in North America. They were those of a small animal about the size of a large cat which had lived in regions which were rich in vegetation and which had many rivers and lakes.

Fig. 15   The scientist constructs the animal from the fossil.(above)

Eohippus, as it has been called, had four-toed front feet, three-toed hind feet and teeth suitable for feeding on vegetation. There appears to have been a long period of time with little or no climatic change and so Eohippus did not change for some 20 million years. Then about 40 million years ago there followed a much drier period which had a considerable effect on the animal and plant life. Many animals became extinct but as many changed slowly to suit the new surroundings which were almost prairie like. Among those which changed was

Eohippus. It gave rise to Mesohippus, a larger animal with three toes on each foot with the middle toe much larger. Its teeth too were much more horse like. Without any doubt, Eohippus and its relatives gave rise to other branches of horse-like animals all of which must have died out as did Protohippus, a descendant of Mesohippus which was very like the present day horse. It was not until about 12 million years ago that an animal which we would recognise as a horse made its appearance. It was called Hipparion. It had feet and legs almost identical with those of a modern horse which did not make its appearance until about 1 million years ago. The only truly wild horse on the earth today is Prezevalski's horse or

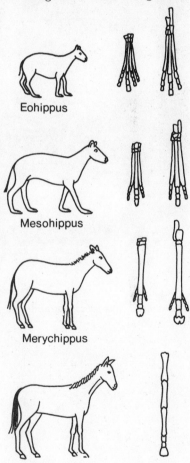

Eohippus

Mesohippus

Merychippus

Horse

Fig. 16   The evolution of the horse.

the   Mongolian   wild   horse.

Many of the so-called wild horses such as the ponies of the New Forest and Dartmoor and the wild horses of North America are domestic horses which have gone wild. Exmoor pones are probably very close to a pure strain descended from the original wild type.

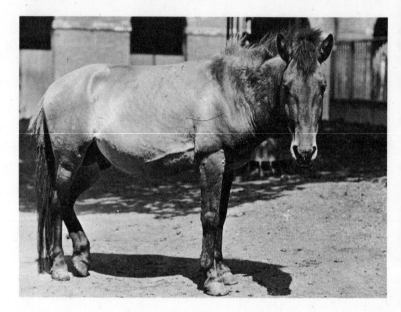

Fig. 17   Prezevalski's horse.

There are, however, many horse-like animals such as zebras and wild asses.

3   The Evolution of Plants

In any discussion on the plant kingdom, the terms "higher plants" and "lower plants" are almost certain to be used. The higher plants are those which have appeared on the earth most recently and whose structure is complicated. The lower plants, on the other hand, are the direct descendants of the earlier plants and usually, are those which have a simple structure. The terms are relative. For example, a moss plant is a "higher" plant than an alga but it is a "lower" plant than a fern. The highest plants of all are the flowering plants. Those of you who have read *Plant Classification (Unit 4)* will also realise that as we move from the lower to the higher plants the reproductive processes and the life cycles change.

All plants have in their life cycle what is called a "gametophyte generation" or "gamete producing generation". This, in most of the lower plants, is the most important part of the plant's life and in many cases it lasts for several years. It ends with fertilisation as a result of which a fertilised female egg cell grows into a "sporophyte" or "spore-producing" generation which is short lived and often it depends entirely on the gametophyte for food and water. In the higher plants, the sporophyte generation is the most prominent one and the gametophyte is small and is sometimes, as in the flowering plants, completely enclosed in the flower.

Let us consider two examples which illustrate this. The liverwort plant is a gametophyte and may live for several years. The sporophyte — the sporangium — on the other hand lives only for a short time and is entirely nourished by the gametophyte. The most prominent stage of a fern is the sporophyte. It produces spores which develop into short lived gametophytes. The evolution of the plants has largely been a case of the gametophyte generation becoming reduced while the sporophyte generation has been enlarged.

The oldest plant remains which have been discovered so far have been algae. These have been found in pre-Cambrian rocks over 1000 million years old.* There are not many fossils but this does not mean that there were not many plants as, if the early algae were

---

* Recent discoveries in some very old sandstones in South Africa seem to indicate that life began at least 3000 million years ago. Fossils found there MIGHT be simple single celled algae.

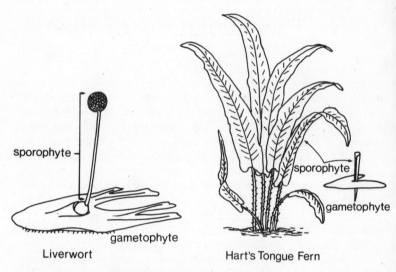

Fig. 18   A liverwort and a fern.

anything like those of the present day they were very soft and so their remains just rotted away. Those which are preserved are those which were encrusted with mineral salts as is Corallina — a common seaweed round our coasts — the others have left no trace.

The first land plants did not appear in the fossil record until the Silurian age about 400 million years ago. Surprisingly, it seems that they were not, as we would expect, liverworts and mosses but were fern-like plants well adapted to live on land. It seems most unlikely that they appeared without some stages to link them with the algae but such stages have not yet been found. The explanation must be that in late Cambrian times the climate was much wetter than it was later and that fossils of land plants which had no tough tissues in them and which should have been the link between the algae and the ferns left no remains.

Fossils of liverworts and mosses do not appear until some 60 million years later in Carboniferous times when the coal measures were produced. There are probably more plant fossils in coal (see Fig. 14) than in any other rock formation. At the time when the coal was being formed vast areas of the earth were covered with fern-like plants and club mosses. The forest floors were wet if not swampy and all plant remains fell into the water. This soon made the water of such a chemical nature that the remains of the plants were preserved instead of rotting away. The ever increasing weight of falling vegetation pressed more and more heavily on the mass of material on the ground so that it was compressed and became coal. Animal fossils are not common in coal possibly because they lived, not in the dark, damp forests but in the drier areas.

Most of the land plants in Carboniferous times were liverworts, mosses and ferns and only towards the end of this period which lasted about 100 million years did the next step in evolution take place. This was the appearance of the Gymnosperms — the naked-seeded plants — which are represented today by our pine trees and conifers.

The Angiosperms or hidden-seeded plants, those plants which we usually associate with flowers, appeared much later — 80 to 100 million years later. Both these and the Gymnosperms have been very

Fig. 19   The scene in Carboniferous times.

successful and together they form the bulk of the land plants of to-day. They have solved many of the problems concerned with living on dry land surrounded by air. They have tissues in them which conduct water about the plant, which give them strength to stand

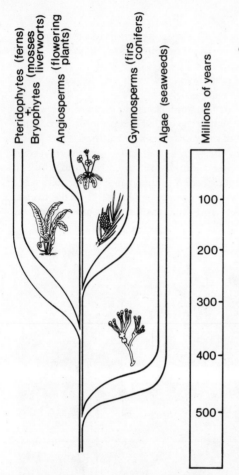

Fig. 20   A time scale
of the evolution of plants.

upright, and which prevent them from drying up. The movement of gametes no longer depends on water and the delicate gametophyte generation is completely enclosed and protected.

This then is a brief outline of the evolution of plants but, before we leave the topic one or two points must be made clear. If we speak of a time when the only vegetation was algae we must not think that the scene was such as might be seen on our beaches and in our coastal waters as the algae of long ago were very different from those we can see now. The vegetation of the Carboniferous times was not in the least like modern vegetation without the flowering plants. The ferns were as big as trees and the mosses and liverworts and club mosses were quite different from their modern descendants.

Another point which must be remembered that once a group of plants or animals becomes established, it never becomes completely extinct. Some types might disappear but the main lines carry on, changed in form and appearance perhaps but still very much with us. Thus, in the case of the plants, we still have algae, bryophytes, pteridophytes, gymnosperms and angiosperms in spite of the fact that they appeared on the earth millions of years ago.

4 The Evolution of the Animals

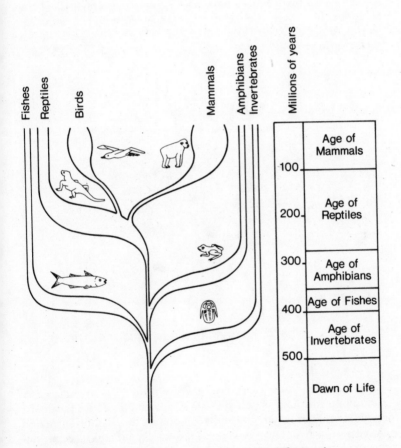

Fig. 21 A time scale of the evolution of animals.

The Earliest Animals

There can be little doubt that the first living things on the earth were
neither plant nor animals but simply small masses of naked protoplasm which, in the course of time, gave rise to plants and animals,
probably in that order. The reason for saying this is that all animal
life depends on plant life as we have already seen in Units 1 and 2.
The earliest living organisms left little behind them to record their
life on the earth but in Precambrian and Cambrian times the seas
were filled with algae among which lived countless single celled
animals not unlike our modern protozoa. Fossil remains have been
found of animals which, like our Radiolaria, had shells which were
almost everlasting. Traces have also been found which show that
evolution has resulted in the appearance of creatures, some like
worms, some like anemones, some like sponges and some like the
shell-bearing molluscs.

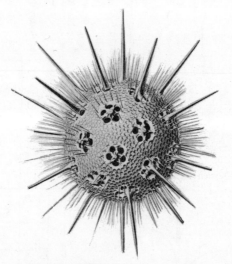

Fig. 22   A drawing of a Radiolarian.

The Age of Trilobites

The climate of those early days was one of cloudy skies and a misty
atmosphere through which the sun's rays could hardly reach the
warm seas. During the 100 million years of the Cambrian period
changes occurred. The seas cooled, the air cleared and it seems that
the amount of oxygen in the air increased. What land there was, was

dry, bare rock and life only existed in the sea where tremendous changes took place among the animals. Many shelled animals appeared, among them those which left "lamp shells", molluscs not unlike cockles, some animals like snails, jelly fish, starfish and sea urchins. The commonest of all were the strange animals called trilobites from the fact that their bodies could be seen to be divided into three parts. These Arthropods invaded all the regions of the seas from the deepest parts to the shallowest and they made up at least half of all the animal population. They,varied in length from 1 cm to 80 cm. Trilobites flourished for some 300 million years but, finally they began to die out and became completely extinct during the Devonian period 400 million years ago. With the trilobites were found some animals the descendants of which, little changed, from their original form, are still found in the mud of the Gulf of Mexico and some of the coasts of Asia. These are the King crabs.

Fig. 23   (left) A Trilobite in Silurian Rock.
(right) An ancient Greek lamp (below) compared with a fossil "lamp shell".

Fig. 24   The marine landscape as it appeared in Cambrian times.

Fig. 25   The King Crab — a survivor from prehistoric times.

The Age of Fish

The next 50 million years, during the Devonian period, saw the real beginning of the vertebrates, the animals with backbones. The first ones were probably quite small aquatic animals with many gill slits used mainly for filtering food out of the water, a simple fin which ran right round the body and a stiff rod of cartilage running from front end to the end of the tail. They might have been similar to our modern lancelets (see Fig. 26). From these evolved the armoured

Fig. 26 The lancelet — a primitive chordate.

fishes. These were curious unfishlike creatures with no jaws and with their bodies protected by large plates of armour. They were probably too heavy to swim properly and simply floundered about at the bottom of shallow water sucking mud into their jawless mouths and extracting their food from it. The sturgeons, garpikes and lampreys still

Fig. 27 One of the early fish — it had no jaws but sucked its food from the mud.

represent them today. The sharks and the bony fish also appeared in Devonian times.

The Animals Invade the Land

The fish which were to be the ancestors of the amphibians also developed in Silurian times. They were like the present day Coelacanths and lungfish. These both have bladders in their bodies by means of which they can take in air and, in the case of the lungfish, the bladders are used for breathing when the water in which they normally live dries up as it does quite frequently. During this dry time they live buried in the mud, often for long periods of time. Both the Coelacanth and the lungfish have fins which are not in the least like the "ray-fin" of the fish with which we are familiar. They are called "lobe-fins" and they have a thick fleshy part which sticks out from the animal's body on the end of which is the fin-like part. The fins are arranged in pairs and it is easy to imagine their owners beginning to walk.

In Devonian times the weather seems to have become warm and dry instead of damp. Ponds and streams began to dry up and the fish had to find new ponds or die. The ray-finned fish died because their fins would not support their bodies on land but the lobe-finned fish were able to move about on the mud by using their fins as legs. From such fish the amphibians developed as animals living mainly on land but having to return to the water to breed. Their skins had to be moist, their eggs were covered with a soft jelly which, if the eggs were not in the water, dried up and the eggs died. The young went through a fish-like tadpole stage. The idea that the amphibians were descended from these fish was at first only a theory but the theory was confirmed when a Coelacanth fossil 350 million years old, which had fins with feet on them, was found some years ago in Greenland.

Fig. 28   A living fossil — the Coelacanth.

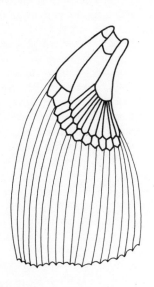

Ray Fin   Lobe Fin

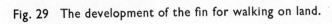

Fig. 29   The development of the fin for walking on land.

The amphibia having progressed to living at least partly on land hardly changed at all for the next 80 to 90 million years. Naturally, their legs became stronger and the animals generally became better suited for their life on land. New types appeared, some of them quite large, but they were all amphibians which were dependent on water and forced to return to it to breed.

The Devonian period seems to have been the "birth age" of many new plants and animals. The 50 million years of this period saw the covering of the land with plants and the appearance of many land animals. Of these, the insects were the most numerous. At first they were very small animals which could not fly but later, in the Carboniferous times, they developed wings and increased in size. Fossil evidence shows us that there were at least a thousand different sorts of insect at this time and that some of them had a wing span of $\frac{3}{4}$ m.

Fig. 30   The Devonian scene.

The Age of Reptiles

It was towards the end of the Carboniferous times that the reptiles, the first real land animals made their appearance. The most important features that they developed were a dry, scaly skin which prevented their bodies from drying up. Fertilisation of the eggs was inside the female body, not external as in the amphibia. The eggs had tough shells and the young which hatched from them were active miniatures of their parents. The reptiles, having severed all connections with the water as an essential part of their lives, spread over the surface of the land. They were of all sizes from about 15 cm to 30 m long. Some which walked on two legs, stood 15 m high. There were among them herbivores, carnivores and omnivores.

Some lived entirely on the land, some flew in the air and some returned to the water to become aquatic reptiles with fish-like features. Some lumbered along on all fours while others ran swiftly on two legs, the front pair being used to hold their prey or, in the case of the plant-eaters to pull down branches of trees.

Among the flesh-eaters Tyrannosaurus was probably the most formidable. It stood 15 m high, it could run swiftly on its powerful hind legs and it had vicious 15 cm long, dagger-like teeth with which it could easily overcome most of the other reptiles on which it fed. The flying reptiles, the Pterosaurs, were curious almost bat-like creatures which had, in some cases, a wing span of 7–8 m. The wings were leathery outgrowths of the sides of the body which were made into wings by being supported by very long little fingers. The other

Fig. 31   Tyrannosaurus rex.

fingers were all free and could be used in a normal way. Judging by the body structure it seems most likely that their flight was mainly a form of gliding rather than true flying. In spite of their success which lasted for some 70 million years, the reptiles were doomed and their numbers began to diminish. By the end of the Cretaceous period, about 70 million years ago, most of them had vanished. It is too much to hope that in some hidden corner of the earth, some of the old types still live just as the Coelacanth has lived in the South African seas for 60 million years. We will only know them as fossils.

The cause of the extinction of these large reptiles is a mystery. Did their bodies outgrow their brains? Did a change in the climate wipe them out? Did some mysterious disease kill them in large numbers? If any one of these is true, why are there still reptiles today? We are never likely to know the true answer and can only guess.

Fig. 32 A life-size model of a Brontosaurus.

The Birds

At about the time when the reptiles were becoming well established, there appeared, as developments from the reptiles, creatures which were to give us our birds and others which were to be the ancestors of the mammals. The earliest known "bird" now called Archaeopteryx, was discovered as a fossil in 1861 and a second one in 1872, both in Germany. They were neither birds nor reptiles but they had features of both. Archaeopteryx was bird-like in that it had feathers, a beak and could fly. It was reptile-like in that it had teeth, a long flexible tail and clawed toes on the legs and on the wings. The brain too was very like that of a reptile. For some reason, after these discoveries of the early "birds" which were thought to have lived during the Jurassic period, no more seem to have lived until late in the Cretaceous period (about 80 million years ago). In rocks of this period fossils of two water fowls were discovered in America. One of them must have been a good flier while the other had very small wings and could not possibly have been able to fly. They both had teeth in their beaks but were much more like birds than was Archaeopteryx.

Fig. 33   Archaeopteryx — the first "bird".

After what seems to have been a period of little evolutionary progress, birds made great advances during the Tertiary period which began about 70 million years ago. As they were the only animals other than the insects which could fly, they were able to escape from enemies and to be able to nest in safety. Their feeding and breeding grounds were considerably extended and they were able to move easily from unsuitable climates to more suitable ones as they do during migration. These advantages over their rivals meant that, at a very early date many kinds of birds, some very like their modern relatives, had appeared. Some of the early birds gave up flight in spite of all its advantages. Many of these were the giants of the bird world reaching a height of 3 m. One of them, the 4 m high Moa of New Zealand, would still have been with us had it not been for man who found it an easy prey. They probably still existed when Captain Cook visited New Zealand in the eighteenth century. The direct descendants of these flightless birds are still to be seen in the ostriches of Africa and the emus of New Zealand and Australia. The well known dodo was also one of the flightless birds made extinct by man, the last one was killed in 1681.

Fig. 34   The Moa — a flightless bird which was over 4 m tall.

One factor which partly accounted for the success of the birds was their possession of feathers which, not only enabled them to fly but, because of their heat insulating properties, allowed them to become warm-blooded and thus more or less independent of climate unlike their reptile ancestors which were cold-blooded. Warm-bloodedness may also have had some effect on the development of the brain as, as a general rule, warm-blooded animals have better developed brains and show more intelligence than do those which are cold-blooded.

The Appearance of the Mammals

The first signs of any animals with any features like those of the mammals were reptiles, the fossils of which occur in rocks of the Permian period. They, the Therapsida, probably existed about 250 million years ago. Among their mammalian features were teeth which were divided into incisors, canines, and molars and, most important of all, it seems certain that their young were born alive.

Fig. 35 The Dodo — a flightless bird which lived on the island of Mauritius until about 1680.

Fig. 36   Reptiles like these were the ancestors of the mammals.

Just how they gave rise to mammals is not known but, somehow, during the next 100 million years, hair appeared, replacing scales, and the animals became warm-blooded. This is known to be true because in the Jurassic rocks, 180–130 million years old, fossils of real mammals appear.

In the development of mammals without doubt one of the most important facts is that their young are born alive and active. This development did not take place in one step. The first mammals laid eggs but suckled their young. The last remaining examples of these are the spiny ant-eater and the duck-billed platypus. The next stage was the marsupials which gave birth to tiny, ill-developed young which completed their development and were suckled in a pouch. These are still found in large numbers in Australia. The final stage was the full development of the young in the mother's body to which it was attached by the placenta through which it was nourished. This meant that the parent was mobile while the young were developing. The mere fact that the young were suckled is important because it means that the parents did not and do not have to seek for food for their young. Another important fact is that, with the increasing certainty that the young will be successfully reared, a reduction of the number of eggs produced has been possible.

Fig. 37   A duck-billed Platypus — a mammal which lays eggs.

The development of the layer of hair which insulates the animals against loss of heat, was also of very great importance as, just as in the case of the birds, it enabled them to occupy areas with widely varying climates. Reptiles and the groups below the reptiles are influenced tremendously by climate, particularly by temperature, whereas mammals are not and are thus found in all parts of the world from the hottest to the coldest.

The success of mammals in differing surroundings has also been helped by the changing of the form of the body to suit varying surroundings. The changes which have taken place apply especially to the feet and legs so that they have become suitable for almost every possible situation.

It is almost certain that the first placental mammals (see above) were small insect-eating animals like shrews and that from this unlikely source all the others including man have developed. It is strange and inexplicable that during the course of their development, as with most other groups, some giant types have appeared, lived for a time and then died out. In the Eocene period 40–70 million years ago there were some real giants. Andrewsarchium was a carnivore 4 m long, Arsinotherium was very much bigger, like a modern elephant and, largest of all, Baluchitherium, a hornless rhinoceros, was 5 m high at the shoulder.

This whole book and indeed one very much larger could be entirely given over to the story of the evolution of the mammals but as it is easy to see the final results of evolution around us and in zoological gardens, the remainder of the section on "Evolution" will be devoted to an outline of the evolution of Man.

The Evolution of Man

Although we usually like to think of man as being something special in the animal world, there is no reason to assume that he is not the product of evolution as is every other animal on earth. As we have already said it is probable that all the mammals originated from small, insect-eating animals like shrews. From these early types which lived about 50 million years ago came the tree shrews and the tarsiers. From these came the lemurs — the ancestors of the monkeys — and it is thought a line of development which led to man.

The tree of evolution of man is very difficult to follow because, as with other animals, there have been many side branches which became extinct while the main trunk continued to develop. Fossils too have so far been scarce and although at present the picture is confused and incomplete, it is certain that in time the whole of man's story will be known.

The story so far began in 1856 near Dusseldorf in Germany with the discovery of a skull which was not quite that of an ape. After much argument, it was agreed by the scientists of the time, that it was the skull of an ape with some man-like features. The skull was given the name of the place where it was found and the animal from which it had come was called Neanderthal man. It is now thought that it is one of the members of a blind branch which ended about 70 thousand years ago. The next discovery was the reward of determined efforts by a Dutch army surgeon named Dubois who, in Java, unearthed a skull and later a thigh bone which must have belonged to an ape-like creature with some man-like features one of which was that it had walked upright. It was given the name Pithecantropus erectus or, more commonly, Java man.

In the 1920s and 1930s excavations were carried out in China as a result of the discovery, in a Chinese chemist's shop, of fossil teeth which were very human in appearance. They were actually being sold as dragons' teeth to be ground up and used in certain magical remedies. Enquiries as to their origin resulted in excavations being carried out near Peking in which several important discoveries were made. These included a complete skull, many bones and thousands

Fig. 38   A tree shrew, (top left) a tarsier (top right) and a lemur.

of stone implements. Peking man was very similar to Java man but he had used tools and knew the value of fire. By 1941, the remains of over forty Peking men had been found but unfortunately all were lost during the war (1941–1946) when they were being moved to a safe place.

Another great source of man's remains is South Africa and work being done there now may provide all the missing answers. In 1924, a skull, so small that it must have been that of a "child", was found at Taungs in the Transvaal. It was buried in deposits thought to be about 2 million years old. In total, three sets of remains were found each with different animal remains which showed that the finds covered the period from 2 million to 800 thousand years ago. Just after the last war, the best specimen of all was found, that of an adult female Transvaal man which was almost "the missing link" which had been sought for so long.

In 1959, Dr. Leakey, a prominent man in the search for man's ancestors, discovered at Zinj in the Transvaal the remains of a creature thought to be 25 million years old. The skull had a chin and it lacked the ape's eyebrow ridges. With it was a stone axe. These remains were thought, as were those of Neanderthal man, to be those of a side branch of the main trunk. There is no way of knowing for certain, as in the rocks of the next 20 million years there do not seem to be any fossils which help us.

We can next take up the story with the finding, again by Dr. Leakey in Africa, the remains of ape-men very like Transvaal man and Java man. From quite numerous remains of similar skeletons, it seems that, with only very small changes, this early type of man existed until a mere $\frac{3}{4}$ million years ago. As far as can be judged they must have been strange creatures. About $1\frac{1}{2}$ m tall and probably very hairy, they must have walked erect with their heads jutting forward. Their teeth were man-like but very large and powerful. They probably used stones, deer antlers and tree branches as weapons and they may also have used fire.

The order in which the fossils have been discovered has not been the order in which they existed as living creatures. This means that we cannot easily follow the correct line. Neither are we sure that some of the fossils are not those of some of the side branches from the main trunk. But, no matter which line he followed, man did evolve until finally modern man arrived on the scene some 50–70 thousand years ago. At first, he lived in small family units in caves and hunted his prey with well-made weapons. Some of them were artists of considerable ability and they left examples of their paintings on the

Fig. 39 Neanderthal man.

walls of caves in France and Spain. They believed in life after death and buried their dead in elaborate graves into which were placed the bodies with the weapons and ornaments used during life.

One of the branches which might be mentioned is that which led to Giant man. As mentioned earlier, many groups have had their giant forms which became extinct and man is one of them. In 1940, there was found, in Java, part of an immense lower jaw which was at least twice the size of any fossil jaw previously discovered. By careful comparison with other specimens and with the jaws of apes, it was decided that it had been the jaw of a "man" very much larger than a gorilla and probably $2\frac{1}{2}$–3 m tall. These giants — there must have been more than one — were not local freaks. A similar but even larger jaw was found in China which was half as big again as the one found in Java which means that its owner could have been $4\frac{1}{2}$ m tall. In Africa also, discoveries have been made of the jaws and teeth of these giants.

Fig. 40　Wall paintings from the caves at Lascaux.

5 Heredity and Mendelism

Evolution depends on the characteristics of one generation being inherited from the one before. This may appear to be a haphazard happening in which the offspring turn out to be like one or other of the parents but there are scientific laws which govern this process of heredity.

The first person to be aware of this was Abbé Mendel who lived in Czechoslovakia in the middle of the nineteenth century. He experimented with garden peas and his first experiments were concerned with the height of plants. Whether by accident or design he was fortunate in his choice of plants because the flowers of the garden pea do not open and are self pollinated. Thus a tall plant is always pollinated with "tall" pollen. If he needed to have a tall plant pollinated with "short" pollen all he had to do was to remove the stamens from the flower of the tall plant so that the only pollen used was that collected from a short plant and placed on the stigmas of the tall plants by Mendel.

From an experiment like this, in which he crossed a tall and a short plant, Mendel discovered that all the plants making up the first generation (called the first filial or F_1 generation) were tall — the short feature had apparently disappeared. But, when the tall plants of the F_1 generation were allowed to produce seeds, the seeds grew into both tall and short plants always in the same ratio of 3 tall to 1 short. The proportion is of course only seen if many plants are used — Mendel grew as many as 8000 plants at once. The fact that the feature of shortness reappeared in the F_2 generation after disappearing in the F_1 generation showed that whatever caused it must have been present in the plants throughout.

In sexual reproduction, the important part of each of the gametes is the nucleus so that it must be correct to say that the factor which governs height must be in the nucleus of, in this case, the pollen grains and the ovules.

Let us now think of this problem of the inheritance of height from this point of view. Think of a tall plant which when allowed to produce seeds produces only "tall" seeds—a so-called pure-bred plant.

The seeds of such a plant must have received a height factor from both the ovule and the pollen grain. If we represent this factor by the letter T then the female gamete (the ovule) can be shown as Ⓣ and the male gamete (the pollen grain) as Ⓣ . The result of the fusion of the two which takes place in fertilisation can thus be shown as ⓉⓉ as every plant must have one factor from each parent for each characteristic.

Now, what happens if the "tall" ovule is fertilised by a "short" pollen grain which can be shown as ⓣ. The plants produced by the seeds of such a union would be Ⓣⓣ. But, these all grow into tall plants which means that ⓣ is overshadowed or dominated by Ⓣ.

In the same way we can show why ⓣ appears in the next generation.

Ⓣⓣ x Ⓣⓣ   The parent plants of the F₂ generation

Ⓣ ⓣ   Ⓣ ⓣ   Pollen grains and ovules carry either
                   Ⓣ or ⓣ

In the pollination, a Ⓣ ovule has an equal chance of being fertilised by a Ⓣ or a ⓣ pollen grain. A ⓣ ovule is the same so that the seeds can be

ⓉⓉ or Ⓣⓣ or ⓣⓉ or ⓣⓣ .

The first three types of seeds would grow into tall plants while the last one would be a short plant because tallness is the dominant character and shortness is the recessive character.

Some characters do not dominate others in this way but work equally to produce an intermediate. Such a case is seen in some primroses of which there are red-flowered and white-coloured varieties. A cross between the two results in a pink-flowered generation. A cross between two pink-flowered plants produces plants with red flowers, plants with pink flowers and plants with white flowers in the ratio one red, two pink and one white.

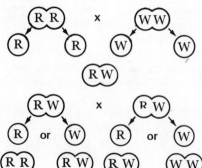

R R × W W	original parents
R R W W	gametes
R W	All F₁ are like this—pink
R W × R W	pink parents
R or W R or W	gametes
R R R W R W W W	2nd (F₂) generation

The way in which characters pair off in this way can be shown by using two well-shuffled packs of cards. One "player's" cards represent pollen grains and the other's ovules. Each player plays one card at a time and the colouring of the pairs is noted. They will come out very nearly 13 red/red, 26 red/black and 13 black/black.

Obviously experiments with plants take many years to carry out as there is only one generation each year and so if work on heredity is to be done quickly some plant or animal with a faster breeding rate must be used. Two such animals are the fruit fly Drosophila and the flour beetle Tribolium. Drosophila matures in 10 days and has many easily recognisable features. The features of Trilobium are more difficult to recognise and it takes 40 days to mature — it is, however, much more easily cared for in the laboratory.

There are many books which give full instructions on how to deal with both these animals and suppliers of zoological materials also usually send full details on how to care for them when stocks are purchased.

6 Cell Division and Heredity

All multicellular plants and animals are produced from a single cell by the nucleus (and the cells) dividing again and again. The dividing of the nucleus is not a simple splitting into two as one might split a lump of putty but a definite process called *mitosis* which has a series of definite steps which never varies.

Mitosis

The nucleus of a resting cell appears to be of an even consistency throughout but, when the cell is about to divide a number of thread-like structures called *chromosomes* appear in it. Soon after they appear, the membrane round the nucleus disappears and the chromosomes move to lie on the equator of a newly-formed structure called the *spindle*. This is like two cones with their bases together and it consists of a jelly-like material. Each cone is composed of a number of smaller cones, one to each chromosome, called *spindle elements*. At this stage, the fact that each chromosome is a double thread is easily seen. The halves of each chromosome now separate and pass to the opposite ends of the spindle. To make sure that they separate completely the spindle also separates into two by the two cones moving apart. The result of this so far is that we have two separate but identical bundles of chromosomes each of which is the same number of chromosomes as there were in the original cell. Each chromosome is also an exact half of one of the original chromosomes. A new membrane forms round each new nucleus and the cell itself divides into two.

The most important thing to remember about mitosis is that both cells produced are identical in every way and are identical with the original cell. Repeated divisions of this sort result in the production of a multicellular organism in which all the cells are exactly the same as far as their nuclear structure is concerned. The number of chromosomes is always the same for any species of animal or plant. In man it is 46, in the rabbit 44 and in Drosophila mentioned above, it is 8. This number is called the *diploid number*.

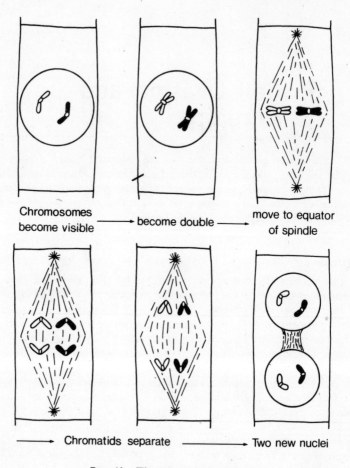

Chromosomes
become visible ——→ become double ——→ move to equator
                                                                     of spindle

——→ Chromatids separate ————→ Two new nuclei

Fig. 41   The stages in mitosis.

Meiosis

If the male and female gametes (e.g. sperms and ova in an animal) were produced by divisions of this sort, they would have the same number of chromosomes as the normal body cells and each generation, caused by the fusion of a sperm and an egg cell, would have twice the number of chromosomes in its cells as had its parents. This would soon result in animals and plants with ridiculous numbers of chromosomes. For example, a man's children would have 96 chromosomes in their cells and their children would have 192.

This is obviously an impossible state of affairs. However, it does not occur because a special type of division takes place during the formation of germ cells. It is called *meiosis* or *reduction division* because the result is that the sex cells contain half the original number of chromosomes — this number is called the *haploid* or *monoploid* number.

Meiosis begins in the same way as mitosis by the chromosomes becoming visible in the nuclei of the cells but as single threads not double ones. Of these chromosomes, half originated from the female parent of the individual concerned and half from the male (see p. 47). These join up in pairs, one "male" to one "female". Thus in Drosophila which has eight chromosomes there will be four pairs.

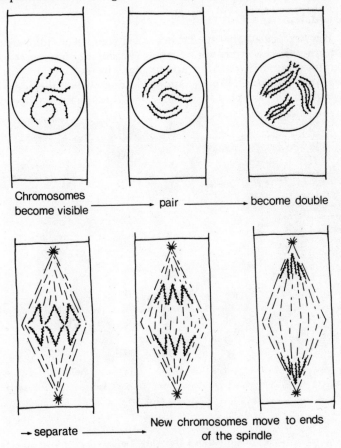

Chromosomes
become visible ⟶ pair ⟶ become double

⟶ separate ⟶ New chromosomes move to ends
of the spindle

Fig. 42   The stages in Meiosis.

These pairs gather on the equator of a spindle which has been formed in the same way as in mitosis and they then split up so that half pass to one end of the spindle and half to the other. This does not mean that all the "female" chromosomes go to one end and all the "male" to the other. Their movement is quite haphazard. In Drosophila, this would mean that this separation of chromosomes results in two cells each with four chromosomes. These new cells divide again in a way similar to mitosis so that we finish with four cells each with four chromosomes. These cells are the gametes. Four is the haploid or monoploid number.

Mendelism and Cell Division

It is now known that the factors called *genes* which influence and control the characteristics of an organism are situated on the chromosomes and that they are arranged in the same way as the beads on a necklace. We also know that in a nucleus there must be two genes for each characteristic — one from the male parent and

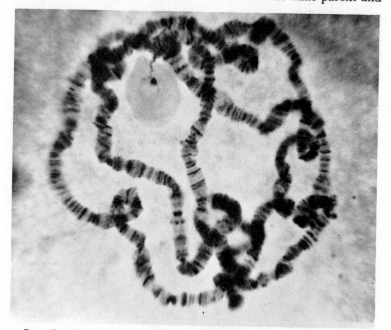

Fig. 43 Giant chromosomes showing bands which *may* coincide with the genes.

one from the female. Bearing these points in mind we can explain Mendel's discoveries (p. 47) in terms of chromosomes and cell division.

In mitosis because each chromosome divides into two lengthwise, each gene is also divided into two so that each daughter cell has the same genes as the parent cell. It therefore has the same characteristics.

In the production of gametes, not only do the chromosomes pair up but the genes are also arranged in pairs. In the first division, the one in which the chromosome number is halved, the pairs of chromosomes and the pairs of genes are split so that each new nucleus will have only one gene for each characteristic. Let us re-examine the way in which tallness and shortness are handed on. The F_1 generation plants are the result of the fusion of gametes which contain genes for tallness on the one hand and shortness on the other — these we have called (T) and (t). Every cell of the plants will contain the genes (T) and (t) which during the meiosis which leads to gamete production, first of all pair up and then separate so that the gametes will have either (T) or (t) but not both. In the fertilisation which follows gamete production, a (T) can pair with a (T) or with a (t), or a (t) can pair with a (T) or with a (t). Thus the plants of the F_2 generation are either

(T T) or (T t) or (t T) or (t t).

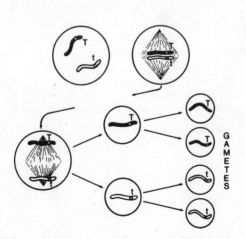

Fig. 44 Meiosis showing how (T) and (t) behave.

Mutations

Sometimes plants or animals appear which have new or different characteristics. These are caused by changes in the genes called mutations. If a mutation results in an individual which is at a disadvantage compared with its neighbours then, in the struggle for existence, it will probably not survive and the mutation will die out. If on the other hand the mutation gives the plant or animal an advantage over its competitors, then the mutation will carry on and in time a new variety will result. Plant and animal breeders are always on the lookout for mutations which have resulted in a new desirable type. By doing this, and breeding from the new types, new varieties of wheat, hornless cattle and the many breeds of dogs, to quote just three examples, have been produced.

It is not certain what produces mutations in nature but it is known that they can be caused by exposure to X-rays and to other forms of radiation and by treatment with some chemicals.

Boy or Girl?

As with other characteristics, the sex of an individual is determined by genes. There are special sex chromosomes which carry the sex genes. They are called the X and Y chromosomes. In all mammals, the female has two X chromosomes and the male has one X and one Y. In reproduction, the females produce ova which carry a single X chromosome while the males produce sperms which carry either an X or a Y chromosome. If a sperm with an X chromosome fertilises an ovum the result will be a female. If the sperm with a Y chromosome fertilises an ovum then the new individual will be a male. Many animals have this arrangement but in birds, the male is XX and the female XY.

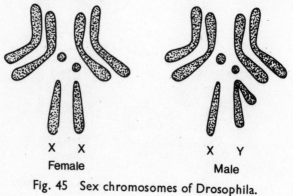

X   X   X   Y
Female   Male

Fig. 45 Sex chromosomes of Drosophila.

Sex Linked Characters

Characteristics which appear most frequently in one or other of the sexes are called sex linked characteristics and, in many animals, the gene for the characteristic is found only on the X chromosome. One common sex linked characteristic is the red/green colour blindness of man. The gene for this colour vision is on the X chromosome and is recessive to the gene for normal colour vision. If a man has the recessive gene on his X chromosome he will be colour blind as there is no gene at all on the Y chromosome. A woman has to have two recessive genes on both her X chromosomes in order to be colour blind. This is unlikely to happen and in actual fact the ratio of colour blind men to women is about 10 to 1. Women can be carriers if they have one X chromosome with the recessive gene and one with a normal gene. She could hand on the gene to a son in spite of not suffering from the complaint herself. Haemophilia — the bleeding disease — is another well-known human sex linked characteristic. The effects of such characteristics can be useful. For example, some breeds of fowls show distinct differences between the male and females at a very early age and make their separation an easy task. Another example is that tortoise-shell cats are always females.

Fig. 46 Sex linked day old chicks. The pullets are dark in colour.

Plant and Animal Breeding

In his search for better animals and plants man has applied the principles of heredity for a long time. Our farms and gardens are stocked with plants and animals which have been produced by the careful selection of breeding animals and plants which have been found to have desirable qualities. Roses in their vast variety were all originally like the wild rose and by choosing plants with different features and by cross pollinating the flowers man has produced all the fine flowers we now have. We also have wheat which will grow in a great variety of climates and are resistant to disease, cattle which are bred as milk producers, cattle which are meat producers, sheep noted for their wool production and so on. In fact nearly every plant and animal which man keeps is the result of careful selection and equally careful breeding.

Fig. 47 Two modern Pigs and their century ancestor, who was about half their size

The Nature of Chromosomes and Genes

We have seen that the features are handed down from generation to generation by genes. This has been known for a long time but it is relatively recently that we have had any idea what they are. We now know that they are made up of molecules of a substance called deoxyribose nucleic acid known for convenience as D.N.A. These molecules have been shown to have a ladder-like form but, a ladder which has been twisted into a spiral. It could also be likened to a twisted zip fastener. The remarkable thing about this molecule is that it can duplicate itself by a process called replication in which the zip — if we are using that comparison — becomes unzipped and each single strand builds itself into a new double strand by the addition of the missing parts. The dividing molecule looks like a letter Y. When the division is complete we have two new D.N.A. molecules where we had one before.

How the Gene Affects the Cell

The characteristics of the cell and thus of the whole organism depend on the protoplasm of which it is composed. The nature of the protoplasm depends on the proteins which it contains. The D.N.A. molecule as well as being able to replicate itself can also produce molecules of ribose nucleic acid — R.N.A. These can leave the nucleus and pass into the cytoplasm of the cell where they act as moulds on which amino acids — the building bricks of protein — are placed to make the proteins of the cell. Thus the D.N.A. controls the R.N.A. which controls the proteins which determine the nature of the cell and thus of the whole organism.

Index